COMPROMISING POSITIONS

The Military as Boss

How'd you like it if your boss dictated where you could go hang out after your work day ends? What clothes you can wear when you're off the clock? Who you could be friends with? Who you could date? Who you could marry? And if, on top of all that, they treated you like a cheat and a liar? Say hello to the U.S. armed forces…

COMPROMISING POSITIONS

COMPROMISING POSITIONS
The Erosion of Trust in the American Military

By

S.A. Wilson

COMPROMISING POSITIONS

This book is published by I&H Recherche Publishing.

For information, address I&H Recherche Publishing, P.O. Box 1586, Cypress, TX 77410.

ISBN: 978-1-937666-06-4

Printed in the U.S.A.

ACKNOWLEDGMENTS

I would like to thank the following for their help with this book: First and foremost GOD, who has always carried me through the difficult times; My parents, for their love and support; My children, the rearing of whom is my real job; And my wife, who has always encouraged and supported all my endeavors.

COMPROMISING POSITIONS

CAVEAT/DISCLAIMER

Anyone reading this book should be aware of the fact that it was primarily written almost 20 years ago while I was still a young officer in the U.S. military. With that in mind, it essentially reflects my feelings at that juncture, as well as facts - and understandings of the truth - as they were known as that point in time. Needless to say, some things have changed (e.g., "Don't Ask, Don't Tell" has been repealed), but I have not really attempted to update this book in any way or make it more consistent with any views that I currently have or that may be popular at the moment. (To do so would have made me feel like I was creating the literary equivalent of a "Ted Turner Colorized Classic.") In short, it is intended to be a "snapshot in time," and remains intact from the time I wrote it except for a few minor changes.

However, this book should not be taken as - and was never meant to be - an indictment of the men and women who have served this country in uniform. They put their lives on the line every day to protect these freedoms that we enjoy (and that are the envy of the rest of the world). Our nation is truly blessed in more ways that we can count, not just because of the liberties that we enjoy, but because of the price we're willing to pay, the sacrifices we're willing to make in the form of America's sons and daughters, to keep them.

PREFACE

I wish to preface this work by stating a few facts about myself and the opinions reflected herein. I am a low-ranking officer in the American military with only a few years of active duty service under my belt. I have neither combat nor command experience. My military career has consisted solely of two assignments, although my years of service can be extended by four if my time as a cadet is brought into consideration.

I freely admit that my vision of the U.S. military establishment has become tainted to a great degree. This disillusion goes all the way back to my first day as a cadet when I realized that the recruiter had lied to me. At that time, however, I was still young enough and naive enough to believe that the recruiter was a single individual; the System could not be held accountable for the misdeeds perpetrated by all of its minions. However, I quickly modified my viewpoint to become suspicious, although not skeptical, of much of what was supposedly ingrained in me during my development as an officer candidate. It was not until later that my suspicions would turn into blatant mistrust of the entire military system.

This mistrust is a by-product of both the things I've heard and the things I've experienced. At first, of course, I chided myself for having these ridiculous thoughts and attempted to find facts to oppose them. Instead of alleviating my fears, however, I discovered only more to justify and encourage them. I learned, to my dismay, that the military has not only molded a system of mistrust but perpetuates it by actions that imply that members of the armed forces - men and women who have taken vows of honor and are willing to sacrifice their lives - are little more than petty criminals and sneak-thieves who will rob John Q. Public blind at every opportunity.

In this book, I will attempt to examine this mistrust, explaining by example how it exists and how it undermines the very foundations of what the U.S. armed forces are supposed to represent. Once proud and noble, the military system has become a mistrustful and bloated bureaucracy that is not worthy of the soldiers who make up its ranks.

There are those who will look upon this work and say that I have neither the knowledge our experience needed to express a true evaluation of the U.S. military. They will argue that in these discussions I am but an innocent child whose young mind can't quite grasp all of the ramifications of certain actions during these trying times. The argument is not without its merits, but I seem to recall someone once saying that children don't know enough *not* to ask the important questions. It is their very innocence that makes them cut through the chaff and get straight to the heart of the matter. Therefore, if I must be rendered a child, it is in this light that I wish my words to be considered.

Finally, I come to my one true shame in connection with what I have written here. I feel it absolutely imperative that my identity remain secret. It is with great remorse that I do this, but I feel it a necessary evil at this time for several reasons. First and foremost among these is the fact that I am still a member of the U.S. armed forces, and it is quite likely that my opinions printed here can be made to hurt me by way of prosecution under the *Uniform Code of Military Justice.* Although this work is merely my opinion, military justice is universally known to be severely lacking in the area of common sense. Actions that are not considered legal in the civilian community can easily get a member of the military court-martialed, fined, imprisoned, and dishonorably discharged - usually all four, although I forgot to mention reduction in rank. If it were a simple matter of losing my job I would not hesitate, but these other (unfair, in my opinion) actions give me reason for pause. And legally battling the military while a member of its forces is impossible; ask those who've tried. As the maxim goes, you can't fight City Hall.

Another reason regarding my use of a pseudonym is far more simple. Even if no action were officially taken against me, there would still be the day-to-day vindictiveness of my fellow officers. Consider the case of the cop who blows the whistle on corruption in his precinct. Despite the fact that he himself has done nothing wrong, he is regarded as a pariah, even by those of his colleagues who weren't on the take. He is shunned and avoided, and often viewed as a disruptive element by his superiors. This, however, is no reason to avoid doing the right thing, and made up only a small part of my decision to remain anonymous. Again, my primary reason is to avoid prosecution under the UCMJ, for whatever reason.

COMPROMISING POSITIONS

Chapter I

Greater love hath no man than this, that a man lay down his life for his friends.

John 15:13

I once believed that the desire to be a soldier was the most unselfish of acts. It was the willingness to make the ultimate sacrifice in an effort to protect all which we hold dear: our country, our values, our way of life. But what was once a noble gesture, a higher calling, has deteriorated into a pact with the devil. A jewel of priceless value has gradually lost its luster, and for a very simple reason: Trust has vanished.

A soldier was once part of a cohesive unit, the personification of *esprit de corps.* That very soldier is now an outsider within the very system he serves. His every action is suspect and subject to immediate and critical review as well as overly-harsh judgment. Distrust is charging through our armed forces with the strength and power of a rogue elephant, flatly crushing everything in its path.

This distrust has grown into a tree with many roots. It is impossible to find out which was the first, the primary root, but there are certain older ones that are easily identifiable. One of these is the treatment of veterans, and one aspect of their treatment has been in regards to medical service. These are men and women who were promised free medical care for the rest of their lives, and yet they are turned away in droves. Unfortunately, it is only now, in their hour of need, that they see how their loyalty and years of devotion are repaid.

Their situation is analogous to the servant who learns, after his employer's death and years of faithful service, that he is not left in the will as promised. It is not the mere fact that he has been excluded, but the fact that there is no acknowledgment of his service, no validation of the fact that his contributions to the welfare of his master's household made any difference. It is a slap in the face with a glove of mail.

The worst tragedy here is that these are people who have *already* made the sacrifices and paid the price for our liberties. If this is the treatment that they receive as their reward, what can someone such as myself, who has yet to truly be called upon, expect? That is the first question. But like water springing from a hole in a dike, it is not long before a few more cracks are spouting liquid. *Why are they breaking their word? What else have they lied about? How can I protect myself?* At this point, the dam bursts and the inevitable disaster follows.

It is amazing to think that our military leaders never envisioned how such a break with their word would affect the way the System was envisioned. I, for one, can tell you that the initial realization that the System was flexible only when it suited its own purposes was a severe wake-up call to both me and my fellows. Unfortunately, I was still too unworldly to see this as a mere symptom of a greater ill. In my young mind, it merely meant that I would need private health insurance after my time was served. That is one of the gifts as well as severe drawbacks of youth, the ability to see things in only one dimension. It allowed me to address situations in a simple, straight-forward manner, but also kept me from seeing the big picture.

Experienced officers, however, knew then what I would only learn in depth later - that sweeping changes could take place overnight, and there were few "Grandfather Clauses" to protect the innocent and unwary. Even more, there was no way to fight the System. There never has been and probably never will be.

It is quite possible that it was a situation like this that gave birth to the initial mistrust, that of the soldier towards the System. The military breaks its word to those who have been faithful and true to its cause. This gives rise to the individual's desire to protect himself from an intangible, internal foe that he can not directly oppose because it is not a corporeal entity. This inability to question and challenge only serves to foster mistrust and feed the fuel of its flames.

Considering how veterans are treated, it is no wonder that current soldiers look at the System with a jaundiced eye. The armed forces have cast off those who served loyally in the past and treats them with blatant disdain. Such treatment only makes military members feel like pieces of meat, whores in effect. Perhaps worse than whores, because after being so obviously screwed, we don't get paid.

It is not, however, necessary to look only at the present plight of our predecessors to determine that the military's values are deeply flawed. One need look no further than the rules and regulations that active duty members are serving under now to see the truth. The System, in every fiber of its being, indicates that it does not trust us, and has instituted an elaborate and excessive combination of checks and balances to assure us of that fact. It assumes from the start, without reason or justification, that men and women who are supposedly of the finest mettle are in fact dishonest and ignoble. No surprise then that servicemen have adopted a "CYA" ("Cover Your Ass") mentality regarding everything they do. Within the first month, a new troop learns that if he is to last any amount of time in the military he must document everything, because his every action will be suspect, and for his entire military career he will be considered not just a possible criminal, but a potential one as well.

But who is the true culprit here? How is it that we find ourselves at this level, where our own people are under constant scrutiny and derision long before any suspicion of wrongdoing? All evidence points toward the System itself. It embodies and perpetuates mistrust from the first moment it is encountered until we finally pry ourselves loose from its grasp. Trust, however, is an intangible that cannot be used for the sole profit of a single individual (at least not for long); it must be mutually beneficial for all parties involved. If one party refuses to place trust and confidence in another, then that same mistrust is returned and should not be unexpected. The System, by breaking its word and never placing initial trust in its people, has - whether by accident or design - both paved the way and carried us to this point.

When viewed over the long term, however, it seems more deliberate than accidental that we have reached this juncture, for there could be no other outcome. We of the military are like a vehicle that has seen hard use and better days, but whose owner refuses to give us the proper service when needed. Under such circumstances, even a novice will tell you that a breakdown is inevitable. And yet, the System continues to treat us as chattel, slaves that will blindly do its will - even to our own detriment - simply because we have no choice and no power. All power here rests in the hands of the System. The abuses that we suffer should come as no surprise then, because, as we all know, power corrupts...and absolute power corrupts absolutely.

Chapter II

And trust me not at all or all in all.

Alfred, Lord Tennyson

The best things carried to excess are wrong.

Charles Churchill

Most people will never know or understand the commitment it takes to serve in the U.S. military. I say that not to imply that they do not have the requisite qualities, but because it is an experience that must be undergone if it is to be fully comprehended. No amount of explanation can prepare you for the demands that will be made of you mentally and physically, and I would be lying if I did not admit to a certain amount of pride when I don my uniform, knowing that I am part of an elite and selective brotherhood invested with the special trust and confidence of the American people.

Unfortunately, the pride I once felt has been greatly diminished. The trust and confidence that loomed so large in the past has dwindled in size to little more than a pinprick. The System, in its infinite wisdom, has found it necessary to not only to question the integrity of military servicemen, but to openly deny it in not just one but several venues. In doing so, however, it reveals its ability and willingness to go to extreme excesses - all in the name of good - that do little more than insult its troops and force them to compromise their integrity.

Consider for a moment the "random" drug test officers must comply with. Officers commanded to submit urine samples for such tests must do so in the presence of an enlisted person. That is humiliating enough, but not only must the enlisted person be present, they must actually see the officer urinate in the jar. Yes! They must actually witness, the urine sample flow directly from one's genitals into the sample jar! It is a completely humiliating and degrading experience, to say the least. Can you imagine Chrysler's Board of Directors approaching Lee Iacocco (or any other CEO) during his tenure and telling him that they had nothing but praise for his work and yet they wanted him to take a drug test? And that while he submitted his urine sample, a low-ranking employee - possibly the newest hire on the assembly line - would be required to watch him and attest to having visually witnessed him provide his sample? Chrysler would have found itself minus one CEO in very short order. Yet the military insists on this type of treatment, the kind of act that demoralizes troops and undermines the special trust and confidence afforded to military officers. Not only are we suspected of using drugs, but our character and moral fiber are so questionable that we must be visually monitored to make sure we do not cheat.

Former Secretary of State George Shultz faced a similar situation himself. Trying to stop information leaks, several officials insisted that everyone - including Mr. Shultz - take periodic lie detector tests. He vehemently refused, stating that if his honor and integrity were questionable, then he had business *being* Secretary of State and the President should never have appointed him. Furthermore, as long as the office was his, he would not demean or dishonor it (and possibly set a precedent) by allowing the character of the person in that position to be called into question. Needless to say, Mr. Shultz never took a lie detector test.

Unfortunately, military members are not allowed to refuse in this manner. Former Navy Lt. Elizabeth Susan Unger faced a court-martial for refusing to allow an enlisted woman to monitor her while she submitted her urine sample. The System not only assumes we are untrustworthy, but will brook no opposition. Thus, it gives rise to distrust, impotence, and frustration among the corps. Also, despite being "random", the drug test is amazingly selective in choosing who has to submit a sample. I have yet to hear of a colonel or general officer being ordered to submit to this indignity, leaving many of us to wonder whether "leadership by example" is merely a convenient slogan for the powers that be, and fostering the us-against-them mentality prevalent in the armed forces today.

This lack of trust also extends into other areas, such as military vacation time, known as

"Leave." Military members receive 30 days of Leave each year. If you leave the local area of your base (usually a 300-mile radius), you are required to be on Leave. This is true even of the weekend; if you go beyond the local area for the weekend, you must use two days of leave for Saturday and Sunday, even if you will be back at your job Monday morning without having missed a single day of work. To most people, that seems extremely unfair, but it is a choice that military members have accepted and chosen to live with. What we have not chosen to live with, however, is the blatant distrust evident in the Leave process.

When seeking Leave, you must first fill out paperwork that your superior approves (by signature) in order to receive the requested number of days. This can be understood since some method of tracking must be in place, and it provides you with written permission to go on Leave. Upon returning to work, however, you are forced to sign another document stating that you only took the number of days you were allowed. In addition, your superior must verify - by signature as well - that you are not lying.

In brief, the System considers you a liar and a thief whose word cannot be trusted. Worse, some commanding officers have even started calling their subordinates on Leave to verify that they are where they say they are. The practice has even begin to include weekends: commanders are starting to call their people on Saturdays and Sundays to ensure that they have not left the local area.

With all this is mind, it is easy to understand how and why our troops are so demoralized today. The System makes it appear that our people lack integrity, that they are dishonorable and must be carefully watched. The people, in turn, criticize the System and senior military leaders for the erosion of trust and the continuous, unwarranted defamation of character that the lack of trust implies.

In addition to this, the long-term effects of the System's distrust are staggering. When an officer is ordered to take the drug test, he is forced to compromise his integrity by submitting to a testing process that, for all intents and purposes, attacks his honor and character. Moreover, the people who serve under him are given pause to wonder at his character and, instead of acknowledging his authority, spend time trying to determine if he is fit to command. But who can blame them? Who among us is willing to be led into battle by and possibly give our lives for someone of questionable morals and values? And if the top commanders don't show faith in our officers, why should the troops? Yet, the

System continues to paint a picture of officers as men and women who, if left to their own devices, will abuse the public trust. Consequently, officers often complain about the lack of trust, prompting the enlisted corps to question it as well. In the end, morale is low because faith in our leaders is diminished at every level of rank and the special trust and confidence our officers once enjoyed has vanished.

For a new troop, the drug test and Leave policy comprise an unnerving yet realistic introduction to the type of environment he has entered. During his training, he is told time and again of the character and integrity expected of a military member. Once on the job, however, he is reminded through daily routine that he is at all times considered a potential criminal. How many people would hire a person they suspected of being a drug-user or a thief, someone that they did not feel they could trust? None, I am sure. The military should be the same. If you can not trust an officer to live up to standards, then you should never have commissioned that person as an officer in the first place.

With the military practically committing character assassination on its members, there is little wonder that distrust is devouring everyone and everything like an all-consuming fire. Those who have found the courage to fight back, like Lt. Unger, have quickly discovered the words of our vows have little substance and less weight. The System, however, holds us to those words with the tenacity of a bulldog, but only when it suits its purposes. Past military leaders would not have allowed the good names of the Army, Navy, Air Force, and Marines to become dishonored by these current regulations that basically accuse our troops of lacking honor and integrity. I can only laugh at the thought of General George Patton's face upon being told that he would have to submit urine test in full view of an enlisted soldier. Or General Eisenhower being told that someone would have to review his Leave statement to ensure that he wasn't lying. They, and any other military leader of the time, would have been furious, and rightly so. We, too, should be furious. These policies are slanderous attacks against the character of all the men and women who wear the uniform and defend this great nation, and do nothing more than help inspire a growing distrust that will eventually destroy the military from within. Without doubt, Former Army Chief of Staff General Bernard W. Rogers summed it up best in this passage:

"If the American people cannot trust our word, if they cannot rely on our conduct, we can hardly expect them to trust us with the lives of their sons and daughters...If we are to have the confidence of the public - and an officer corps worthy of the name - we must recapture our sense of indignation. We must treat those persons who disgrace our good name with the disapproval they deserve.

"Those within our ranks who disgrace our good name are relatively easy to deal with. They can be brought to trial, stripped of their status as officers, and dismissed from the service. Much more insidious than such rogues and scoundrels, however, are those highly placed governmental functionaries who - in pursuit of what they see as a greater good - unwittingly undermine the special trust and confidence the officer corps traditionally deserves. These bureaucratic bunglers truly should excite our sense of indignation for, left to their own devices, they will surely destroy the very foundations of honorable service."

Chapter III

Woe unto them that decree unrighteous decrees...

Isaiah 10:1

Government is not reason; it is not eloquence; it is force! Like fire, it is a dangerous servant and a fearful master.

George Washington

Understandably, the complexity of relations between individuals in the military increases exponentially when compared to the civilian sector. We operate in a stressful environment, and the thought that a close companion could die makes many of us withdrawn, and military commanders seem to prefer it this way. However, we are human beings, and as such we engage in human behavior. Unfortunately, human behavior is apparently a shortcoming in today's military.

COMPROMISING POSITIONS

Consider the following situation, a portion of which most of us are already familiar with:

Boy meets girl.

Boy and girl fall in love.

Boy is demoted, dishonorably discharged, fined and sent to prison.

To most people, the outcome here can only be described as bizarre, and even that is an understatement. Unfortunately, this is not just some twisted nightmare, but the facts of life in today's military.

Fraternization is the word used to describe an improper relationship (usually a romantic one) between an officer and enlisted person. Falling for the wrong person is one of the quickest ways to not only end a military career, but to also go to prison. Never was the phrase "All's fair in love and war" more profound. It seems very odd that men and women who are given the power to destroy millions of lives can not be trusted to exercise the proper judgment in determining who they shall woo romantically. The System, however, views the situation differently, and an officer who chooses to date an enlisted woman risks his career, reputation, and freedom by flaunting the rules and doing so.

The rules, however, are not steadfast, and it is not always the boy who gets the short end of the stick; just ask Lt. Kelly Flinn, who had the misfortune to fall in love with a married man. Lt. Flinn, a graduate of the Air Force Academy and the nation's first female bomber pilot, found herself charged with a barrage of crimes, including adultery (Lt. Flinn is single), fraternization (the man was a civilian, although his wife was enlisted), and sexual misconduct - a charge that does not even exist in the civilian world.

When questioned about the affair, Lt. Flinn lied under oath regarding her relationship with the man, a soccer coach named Marc Zigo. While neither Lt. Flinn's lying nor Mr. Zigo's infidelity can be condoned, it should have been understood and accepted that both parties, as mature adults, mutually consented to the relationship. It is not the morality of their choice that is really in question, but their right to choose. In the civilian sector, men and women are allowed to have certain inalienable rights which we in the military have been stripped of. Think of your significant other and ask yourself what your reaction would have been had someone told you, without reason, that you were not fit to choose the person with whom to share your life. Yet, the System upholds and enforces such a law, immoral as it may be.

Oddly enough, although the Uniform Code of Military Justice (UCMJ) makes it a crime for an officer and enlisted person to date or be romantically involved, it is perfectly legal for them to be married. This hypocrisy is little more than a cruel joke, however, for an officer who marries an enlisted person can expect to be investigated. Should evidenced turn up to prove that fraternization occurred prior to the marriage, the officer can still be court-martialed and sent to prison, a fact to which many former officers can attest.

Air Force Capt Douglas Chin is another casualty of the fraternization wars. In the fall of 1996, Capt Chin began dating an enlisted woman who would eventually become his wife. Traditionally, fraternization in the military has been allowed as long as the officer was not in the enlisted person's chain of command; in other words, the enlisted person could not work for the officer. Capt Chin and the enlisted woman, Airman 1st Class Melody Fellure, dated discreetly and worked in separate organizations. At the point they decided to marry, A1C Fellure applied for an early separation from the Air Force which was granted. Unfortunately, a complaint about the relationship surfaced, and although Capt Chin never lied about his relationship and obeyed all the traditional rules concerning dating enlisted, he was eventually discharged from the Air Force. The rules, it seemed, had changed without warning.

But drastic rule changes are no longer novel, and relationships in the military are not restricted merely between officer and enlisted. There also exists "unprofessional relationships" which occur between two officers or two enlisted when the disparity in the ranks is too great. To illustrate, an Army Colonel and an Army Captain who are dating can be accused of having an unprofessional relationship. However, the concepts of fraternization and unprofessional relationships go far beyond this. An officer who plays basketball at the base gym had best make sure that the other players are all officers as well; fraternization and unprofessional relationships are not merely limited to male-female relationships, nor just romantic ones. It covers all dealings between officers and enlisted personnel.

It is, in short, frighteningly reminiscent of South African apartheid; we are told who we can date, who we can marry...even who we can be friends with. Under what other system can falling in love - or simply being someone's friend - be considered a crime? The System, however, insists on trying to force us to be unfeeling automatons by maintaining a rigid control over the personal lives of its members.

Unfortunately, that control helps create and perpetuate an environment wherein one must often compromise one's integrity: being sneaky and covert (in essence, fostering distrust) in order to find happiness on a personal level. I know of no military member who wants to lie about his life, yet many are forced to do so in an effort to keep their military career on course. And to spread the misery further, those of us who are aware of fraternization and unprofessional relationships are required by law to betray the friendship and trust of our military brethren by reporting them. The choice is not an easy one: to obey an unjust law, or compromise our integrity and submit to it when we know it is wrong.

In order to help alleviate this type of burden from its people, the military adopted the "Don't ask, don't tell" policy regarding the sexual orientation of its members. In short, you can not ask a military member if he or she is homosexual, and they do not have to respond to such questions when asked. It has been suggested that this policy be expanded to include all relationships. (As of this writing, commanders can ask questions of an extremely personal nature - as they did in the Kelly Flinn case - and you are compelled by law to tell the truth). Even should that happen, the law is still likely to be little protection, for the System has the ability to break its own rules with indemnity. A prime example of this is the case of Air Force Major Debra Meeks. In her case, the Air Force violated its own policy of "Don't ask, don't tell" when agents from the Air Force Office of Special Investigations (OSI) interrogated a civilian woman without cause and on their own initiative until she revealed knowledge of Major Meeks homosexuality.

What this clearly shows is the System's penchant for having its way despite information or rules to the contrary. Who would not be fearful of committing himself totally to and placing his life firmly within the grasp of a despot? And yet, that is what members of the military have done. The System is an unforgiving oligarchy that rules without thought or reason, particularly in the area of one's private life. It comes as no surprise then that, after watching so many of our fellows victimized by the those in the upper echelons of power, those of us still serving place no faith in the System and take every opportunity to protect ourselves. In our minds it is Grendel, waiting only for night to fall in order to descend upon us and shred our lives. It can not be trusted to look out for our interests at any level, and so we are ourselves forced to stoop to its level, play its game and submit to our own role in it. We compromise our integrity and character not for any type of gain, but just to remain on an even keel. The willingness to die for our country should, at the very least, buy us the right to seek just a little bit of happiness by choosing our own friends, loved ones and companions. Unfortunately, our decision to do so puts many of us in the position of having to lie and deceive, thus adding to the ever-developing growth of distrust in the military.

Chapter IV

The rain it raineth on the just

And also on the unjust fella

But chiefly on the just, because

The unjust steals the just's umbrella.

Lord Bowen

Next, it's probably worthwhile to point the spotlight on the military's wasteful fiscal policies. At the end of each fiscal year, those in charge of military resources are commanded - yes, commanded - to spend billions of dollars on unnecessary items. This is because if the money is not spent by the end of the year, it goes back to Congress. Rather than see the money revert back to the government and possibly be used for other things (education, reducing the deficit, etc.), the military spends it on anything and everything it can, all the way down to the last penny. Part of the reason for this is survival. With all of the cutbacks and drawdowns in the last few years, the entire military-industrial complex is fighting for its survival. Every armed services program is afraid of having its budget or manning cut. The logic, therefore, behind spending all money received is this: if any money is given back, the powers that be will assume that your budget is too large and possibly cut the amount of money you receive next year. Therefore, even though you currently have everything you need for your program and an extra $10 million dollars in your account, you'd better spend every penny of it, or face the prospect of possibly receiving only half of what you need to operate next year.

There is also an unanticipated side effect from the excess spending: redistributing old items that are displaced when new items are purchased. For example, suppose Captain Kirk has $1 million in excess funds at the end of the fiscal year. He is ordered to spend it, so he decides to buy 300 new computers for his troops. When the new computers arrive he takes out the old computers and turns them over to a military reutilization office. The reutilization office is supposed to find new homes for the old computers. It sounds simple enough in theory, but that is rarely the case. Reutilization offices often do not find homes for the old equipment, and because of the exorbitant amount of spending and equipment that gets turned in, they often do not even have room to store it. Many facilities have had to resort to storing equipment - desks, chairs, and even almost-new computers and printers - *outside and uncovered* against wind, rain, dust and the elements! Those of us who attended public schools find it extremely hard to believe that there are no school districts that are willing to accept free computers, office equipment, and furniture. The waste of both money and equipment is unbelievable and extremely frustrating to those military members who can not afford a much-needed home PC, but are forced to watch thousands of computers (among other things) rust in the sun.

Chapter V

I repeat...that all power is a trust - that we are accountable for its exercise...

Benjamin Disraeli

Unlimited power is apt to corrupt the minds of those who possess it.

William Pitt, Earl of Chatham

Now, let's address the callous disregard - although "contempt" would not be too strong a word - in which the military holds its members. The System has proven to us time and again that we are little more than cannon fodder, refusing to provide us with the tools necessary to do our jobs. A recent newspaper article highlights this fact with the words of one veteran who notes that as a machine-gun squad leader in 1951, his two guns were M-1918 Brownings. In 1961, after three wars, those same M-1918 machine guns were still being used. In 1965, the new M-60 was issued. Unfortunately, those exact same M-60 machine guns are still being used today, and after three decades of service they have garnered a solid reputation for jamming.

At the same time that our young men are being sent out to fight with antiquated weapons, however, our tax dollars are being squandered on the sumptuous and opulent indulgences and of senior military leaders. For example, one general, upon receiving $5600 in new office furniture, demanded that it all be sent back because it did not suit his taste. According to reports, the general stated the furniture was "...traditional walnut...It's too dark." With military priorities this disorganized, it is easy to see why our people are distrustful of our commanders and more fearful of going to war than ever before.

But even worse than having to confront an external enemy with superior weaponry is the battle we have to fight with the enemy within. Every military member knows that at any time he may be called upon to make the ultimate sacrifice. It is something that we all understand and accept. We are ready and willing to die on the battlefield at the hands of our foes. What we are not willing to do is be intentionally harmed by those we have vowed to protect, and yet that is what has happened.

The Army has only recently admitted to deliberately injecting scores of African-American soldiers with syphilis many years ago. (With mandatory annual vaccinations for influenza and hepatitis, similar events could be happening today) Even more recently, the government has purposely withheld information on Gulf War Syndrome, citing that no more than 100 troops could possibly have been infected by chemical and biological weapons while serving in Iraq. The actual number is now conservatively estimated to be 60,000.

To the general public, these facts are shocking; to military members, they are almost commonplace. After all, it is has been clear for quite some time that the System is not a friend or companion but rather an opportunist who will disregard any services you have rendered to him in the past in order to achieve his own ends. Those of us currently on active duty are not frightened by the revelations of what the military has done to its own in the past. We are far more disturbed by what the future will reveal was done to us during our time in uniform.

Chapter VI

The greater the power, the more dangerous the abuse.

Edmund Burke

Another factor adding to the stress of life in the armed forces is the concept of a "one-mistake" military. What this essentially means is that you have to be perfect, because if you make a single mistake your military career is, for all and intents and purposes, over. And it could be something as simple as a speeding ticket. If word gets back to your commander, you can expect your rear to be flayed, sautéed and flambéed. Bouncing just one check or missing a single payment on your financial obligations could also do it. This rule is especially prevalent in the work place - even on items beyond your control.

For instance, if a contracting officer orders an item for his boss and the contractor does not deliver it on time, no one would be surprised if the contracting officer were harshly reprimanded, despite having done everything right. And excuses for mistakes are sometimes even worse than the mistakes themselves. Those who have excuses - even legitimate ones - are assumed to be incapable of accepting responsibility, and that is the death knoll of any military career. It becomes, therefore, very difficult and exceedingly frustrating to maintain your integrity in an organization where a single misstep can cost you your career.

Military leaders, however, argue that the "one-mistake" military does not exist. They say, that we have never been a one-mistake armed force, but that we have instead always been a "one-crime" armed force. But it is rather convenient for them to use the term "one-crime," because under the UCMJ, *any* action can be considered a crime. "Conduct unbecoming an officer" is a charge that pretty much covers all the bases. Picking your nose in public, spitting on the sidewalk, even forgetting to flush the toilet can all be considered chargeable crimes. In short, if the military wants to get you, they can, and there's nothing you can do about it because the "unbecoming conduct" charge can be anything.

Therefore, few members of the military dare to speak out against any policies and activities that they disagree with because it is far too easy for the System to retaliate. Little wonder then that I choose to write this book under the protection of a pseudonym. And people wonder why Oliver North followed the orders he was given. Plainly stated, even if he were not a loyal soldier following orders, Lt Col North could have done nothing without jeopardizing his own welfare. And the military does not act on anonymous tips. Therefore, anyone who wishes to bring an issue to light must expose himself as well. It's a lot like a witness who wants to testify against the mafia, but in order to do so must tell his antagonists exactly where he will be at all hours of the day. The impotence that many of us feel only serves to drive morale lower and widen the gap between our personal beliefs and mandates of the System we serve.

Chapter VII

Dictators ride to and fro upon tigers which they dare not dismount. And the tigers are getting hungry.

Winston Churchill

Another growing point of contention between the military and its members is the escalating level of influence that the armed forces have over one's personal life. We understand that we can be called to war at any time, and that the nature of our business demands that we comply with strict and rigid standards because lives are on the line. However, we should be able to live our private lives as we see fit. Unfortunately, the military has begun to fill the role of *1984*'s "Big Brother" all too well.

Every facet of our off-duty lives - from the clothes we wear to the hobbies we pursue - are now being dictated by senior military leaders. Only a few short years ago, an Air Force major was reprimanded for wearing an earring during off-duty hours. In another instance, another person was charged with listening to music with crude and explicit lyrics. The military has also declared certain styles of dress to be fads and declared them to be off-limits. One of the clothing items specifically mentioned are the jogging suits worn by many popular African-American rap stars. The intentions of the military here can only be assumed to be good, but they seem to isolate a particular culture and attempt to eradicate it: that of African-Americans.

Rap music (much of which contains explicit lyrics) is banned, as well as jogging suits, which have almost become a cultural style of dress. At the same time, however, Caucasian military members are still permitted to wear "cowboy" outfits - boots, monstrous belt buckles, ten gallon hats - not only on base but in the officer and enlisted clubs as well. The argument here, possibly, is that the cowboy is a part of the heritage of many Americans, and they should not be denied the right to exhibit that. However, I wonder what the reaction would be if an African-American entered a base club wearing an African dashiki, or if an Asian-American entered wearing a Japanese kimono.

Racism, however, is not the point here (that is an issue that I do not delve into very deeply, for it could fill an entire book by itself). Instead, the focus is on how the military has expanded its role to include controlling not only the on-duty actions of its personnel, but their off-duty behavior as well. Military members have, for quite some time, been feeling stifled by this, and their growing disenchantment with being unable to live their lives as GOD intended is chafing to the point insanity.

Chapter VIII

How the war veteran is perceived to be treated will directly affect the willingness of the future generations to make the same sacrifice.

George Washington

Finally, let's re-examine the situation of the 25.6 million veterans and retirees. As previously mentioned, it is here that the strongest argument can be made for distrusting the military. Nowhere else are the blatant lies and outright deception more visible than in the treatment of these men and women, beginning with the promise of free medical care - for life! - for all military retirees. Instead, they receive treatment at military facilities on a "space-available" basis. This is even true of medical facilities that were built specifically for military members and their dependents using servicemen's dollars. For example, the Pensacola Navy Hospital was built using $0.20 deductions from the monthly pay of Navy seamen and Marines. Those same GIs, in retirement, are now being told that treatment for them and their dependents at that hospital is optional.

Imagine making mortgage payments on your dream house for thirty years, and then being told that another family will be living there! And adding insult to injury is the fact the they can receive space-available care only until age 65, after which they are refused treatment. But not only are they refused treatment, they must also pay for a portion of their care under both Medicare and the military health care system known as CHAMPUS (Civilian Health and Medical Program).

In its defense, the government claims that Public Law 569 in 1956 changed full health care benefits from a right to a privilege. Be that as it may, military recruiters and recruitment literature continued making the promise of free health care well into the 1990s. The System, however, has refused to make good on its word to the men and women who have served this country with unwavering devotion.

Consider their plight in comparison to the true story of violent inmates at a Florida prison who were promised early release. These prisoners were promised "gain time" which would reduce the length of their sentences. When the state of Florida tried to renege on this promise, the Supreme Court stepped in, ruling that promises made - even those to violent criminals - must be kept. It is indeed shameful that the most powerful nation on earth is more eager to defend the rights of convicted felons who have broken its laws than those of honorable veterans who have protected its borders.

If this were not enough, dealings with the Veteran's Administration (VA) have been even more frustrating. Prior to 1988, if the VA said that you were not entitled to benefits, there was nothing you could do unless you could convince the Board of Veteran Appeals that the local VA office had made a mistake. After 1988, however, the Court of Veteran Appeals was established and began handing down decisions that forced the VA to change its ways. Thousands of cases were remanded to local Regional Offices to be readjudicated, causing a massive jam-up that still persists to this day.

Even so, the deck still remains stacked solidly against the veteran. A veteran who files a VA claim is prevented by law from getting legal representation until the Board of Veteran Appeals turns down his claim. On the other hand, VA adjudicators are ablr to rely on a local VA counsel to advise them. Regional

Offices get advice from the Office of General Counsel, and the Board of Veteran Appeals has its own lawyers.

Armed with this kind legal talent, it would seem that the VA would, at the very least, be willing to play by the rules. Still, several VA lawyers, having no legal basis on which to deny claims, have been caught destroying evidence in order not to lose cases. All in all, it paints a very grim picture of what future retirees and veterans can expect for their long years of service.

Epilogue

The only thing necessary for evil to triumph is for good men to do nothing.

Edmund Burke

In closing, I hope this book frankly addresses what this distrust between the military and its members means: those of inflexible honor and integrity will be unable to continuing serving, leaving the military in the hands of less capable and less scrupulous individuals. These, in turn, will think that everyone else has their own moral shortcomings and, therefore, institute policies that inspire even greater distrust of the System. This, of course, will only drive out even more of those men and women that the military should be retaining.

In short, it is a vicious circle, and one that will eventually leave our military in the hands of moral degenerates. In some instances, it has already happened, as evidenced by the rape trials of Army drill instructors at Aberdeen Proving Grounds in Maryland. These trials prove, if nothing else, the growing lack of standards and values among the military. However, it is not so much that military standards no longer exist, but rather that the System has driven off all the men and women who would have willingly adhered to them, and is, instead, left with those of questionable character in command.

In retrospect, it is not enough that we adopt the current military philosophy of "do more with less" (meaning that current military members, aside from having to pay more for growing medical and dental costs, have reduced benefits and fewer privileges, but are required to work longer hours, more days, and be deployed longer than ever before). We are also asked to ignore many of the good and decent things within ourselves, and all for a government that has and will continue to break its word to us. The System has betrayed the noble members who make up its ranks, and it is blatantly obvious what the results of that betrayal are: distrust, frustration, and ambivalence. Our conscience, honor, and integrity ask one thing of us; the military's deceit, disdain and disregard for us demand another. The Bible tells us that no one can serve two masters, for he will love one and hate the other. For members of the armed forces, we can either loathe the military or loathe ourselves. Some of us have chosen the former, but far too many have chosen the latter.

Made in the USA
Monee, IL
05 November 2026

67375739R00042